Furthermore, this publication pertains equally to small-unit leaders and senior commanders. Since intelligence is an essential component of any military activity, this manual is meant to guide Marines at all levels of command in both the operating forces and the supporting establishment.

C. C. KRULAK
General, U.S. Marine Corps
Commandant of the Marine Corps

DISTRIBUTION: 142 000005 00

D1303356

Intelligence

Chapter 1. The Nature of Intelligence

How Important Is Intelligence?—The Objectives of
Intelligence—Intelligence as Knowledge—Intelligence
as a Process—Why the Mystery?—What Makes Intelligence
Different?—Expectations of Intelligence—Intelligence in the
Information Age—A Case Study: Desert Storm 1990–1991—
Conclusion

Chapter 2. Intelligence Theory

What Do We Want to Know About the Enemy?—
Characteristics of Good Intelligence—Classes of
Intelligence—Capabilities Versus Intentions—Signals
and Noise—Levels of Intelligence—Intelligence
Requirements—Sources of Intelligence—Functions of
Intelligence—Security—The Intelligence Cycle—A Case
Study: Vietnam 1972—Conclusion

Chapter 3. Creating Effective Intelligence

The Challenge to Intelligence—Intelligence Is a
Command Responsibility—The Command-Intelligence
Connection—The Intelligence-Operations Connection—

Intelligence as a Team Effort—Intelligence Is a Product, Not a Provision—A Balanced Approach—Focusing the Intelligence Effort—Generating Tempo Through Intelligence—Intelligence Education and Training—A Case Study: Somalia 1992–1993—Conclusion

Notes

Chapter 1

The Nature of Intelligence

"And therefore I say: Know the enemy, know yourself; your victory will never be endangered. Know the ground, know the weather; your victory will then be total."[1]

—Sun Tzu, *The Art of War*

"For the whole reason-for-being of all military intelligence personnel is to facilitate accomplishment of the mission, and to save lives. When they fail, all the wrong people are hurt."[2]

—Stedman Chandler and Robert W. Robb, *Front-Line Intelligence*

To develop effective intelligence, we must first understand its fundamental nature—its purpose and characteristics as well as its relationship to command and operations. This understanding will become the basis for developing a theory and practical philosophy for intelligence.

HOW IMPORTANT IS INTELLIGENCE?

Maneuver warfare requires a firm focus on the enemy. It aims at taking action which avoids enemy strengths and exploits enemy critical vulnerabilities. The identification of these strengths and vulnerabilities is crucial. Maneuver warfare requires acting in a manner to deceive and then striking at a time and place which the enemy does not expect and for which he is not prepared. Identification of an adversary's expectations and preparations is also important. Maneuver warfare requires decision and action based on situational awareness—a keen understanding of the essential factors which make each condition unique—rather than on preconceived schemes or techniques. How is this situational awareness gained?

Accurate and timely intelligence—knowledge of the enemy and the surrounding environment—is a prerequisite for success in war. Certainly, maneuver warfare places a heavy emphasis on the judgment of leaders at all levels. Nonetheless, judgment, even genius, cannot substitute for good intelligence. Genius may make better sense of available information, and it may provide superior and faster use of the knowledge it gains from that information, but no command- er—no matter how brillian-t—can operate effectively without good intelligence. A brilliant commander, Field Marshal Erwin Rommel, proclaimed that, "It is not that one general is more brilliant or experienced than the other; it is a question of which general has a better appre-ciation of the battlefield."[3]

Intelligence, therefore, is at once inseparable from both com-mand and operations. Intelligence contributes to the exercise of effective command during military operations and helps ensure the successful conduct of those operations. By identifying en-emy weaknesses susceptible to attack, intelligence also serves as an important element of combat power.

Effective intelligence in the hands of capable commanders has often provided decisive advantages of tactical, operational, and strategic importance. The Battle of Midway in June 1942 was won by a vastly outgunned and outnumbered American fleet because its commanders had received, recognized, and acted upon detailed and accurate intelligence. In 1986, during air strikes conducted in response to Libya's terrorist activity, intelligence provided the detailed understanding of the Libyan air defense system that enabled Marine and Navy aviators to effectively shut it down. Intelligence's identification of critical vulnerabilities in Iraqi air and ground defenses contributed to the rapid and thorough defeat of Iraqi forces during Operation Desert Storm.

THE OBJECTIVES OF INTELLIGENCE

Understanding the relationship between intelligence and command and control is key to understanding the role of intelligence. Command and control is about making and executing decisions. The main purpose of intelligence is to support the decisionmaking process.

Intelligence strives to accomplish two objectives. First, it *provides accurate, timely, and relevant knowledge about the enemy (or potential enemy) and the surrounding environment.* In other words, the primary objective of intelligence is to support decisionmaking by reducing uncertainty about the hostile situation to a reasonable level—recognizing, of course, that the fog of war renders anything close to absolute certainty impossible.

In achieving its primary objective, intelligence performs four related tasks. First, it identifies and evaluates existing conditions and enemy capabilities. Second, based upon those existing conditions and capabilities, it estimates possible enemy courses of action, providing insight into possible future actions. Third, it aids in identifying friendly vulnerabilities the enemy may exploit. Finally, intelligence assists in the development and evaluation of friendly courses of action based on the results of the first three tasks.

The second intelligence objective is that it *assists in protecting friendly forces through counterintelligence.* Counterintelligence includes both active and passive measures intended to deny the enemy valuable information about the friendly situation. Counterintelligence also includes activities related to countering hostile espionage, subversion, and terrorism. Counterintelligence directly supports force protection operations by helping the commander deny intelligence to the enemy and plan appropriate security measures.

The two intelligence objectives demonstrate that intelligence possesses both positive—or exploitative—and protective elements. It uncovers conditions which can be exploited and simultaneously provides warning of enemy actions. Intelligence thus provides the basis for our own actions, both offensive and defensive.

INTELLIGENCE AS KNOWLEDGE

Although the objectives of intelligence have been discussed, the term *intelligence* has not been defined. Very simply, intelligence is knowledge—knowledge about the enemy or the surrounding environment[4] needed to support decisionmaking. Since people understand situations best as images—mental pictures—intelligence aims to create an accurate or meaningful image of the situation confronting a commander. Good intelligence paints a picture—or more accurately, several pictures—of possible realities.[5]

Not all knowledge which goes into military decisionmaking qualifies as intelligence. Knowledge pertaining directly to the friendly situation or to the status of an ally does not constitute intelligence. Knowledge not pertaining directly to the friendly cause generally falls under the category of intelligence.

What do we mean by knowledge? In describing intelligence as knowledge, we are distinguishing intelligence from data or information.[6] Intelligence is developed from information, but it is important to recognize that *intelligence is not simply another term for information.* Information is uneval- uated material of any kind—enemy prisoner of war interro- gation reports, radio intercepts, reconnaissance reports, photographs—and represents the raw material from which intelligence is ultimately derived. Few pieces of information speak conclusively for themselves. They must be combined and compared with other pieces of information, analyzed, evaluated, and, finally, given meaning. Good intelligence does not simply repeat the information which a source reveals. Rather, it develops this raw material in order to tell us what that information means and identifies the implications for decisionmaking.

In other words, *intelligence is the analysis and synthesis of information into knowledge.* The end result is not more information, but knowledge that gives us a meaningful assessment of the situation.[7]

Since intelligence is derived from information, it shares many attributes of information. Information, and the intelligence which results from it, is perishable. Information will always be incomplete, sometimes confusing, and often contradictory. Not all information will be important or even relevant, and much of it may be inaccurate or misleading. Too much information can be as harmful as too little. With all information, we seek not a large amount, but to have the right information available when needed and in a useful form, and so it is with intelligence.

Finally, we note that knowledge does not exist for its own sake, but as the basis for action. We do not develop lengthy intelligence studies just because we have the ability to do so or because a subject is of academic interest. Intelligence that is not acted upon or that does not provide the potential for future action is useless.[8]

INTELLIGENCE AS A PROCESS

Intelligence should be thought of as not just a product—knowledge—but also the *process* which produces that knowledge. Intelligence is the process which identifies and evaluates existing conditions and capabilities, estimates possible enemy courses of action based upon these conditions and capabilities, and assists in the development and evalu- ation of friendly courses of action—all in support of the commander's decisionmaking.

Intelligence is thus a central component of the command and control process, which can be described by a simple model known as the observation-orientation-decision-action (OODA) loop. Intelligence activities make up a significant portion of the observation-orientation phases of the OODA loop with the primary purpose of supporting the decision phase. Intelligence also supports the action phase by identifying targets for attack and by assessing results, bringing the OODA loop full circle to the next observation phase in support of a subsequent decision.

Intelligence must not be construed as the exclusive province of intelligence specialists. Intelligence activities are driv- en by the need to answer questions crucial to the planning and execution of operations. Intelligence is inseparable from operations. Data collected during the course of operations is essential to the development of a timely and accurate intelligence picture. Above all, intelligence shapes (some would say drives) the decisions made during the conduct of operations. *All Marines involved in operations are involved in intelligence in one way or another, and all Marines involved in intelligence are involved in operations.*

WHY THE MYSTERY?

In the past, there has been a perception that intelligence is a highly specialized field shrouded in secrecy and isolated from other warfighting areas. Many misconceptions concerning intelligence have arisen; some even view it as the modern equivalent of wizardry. Why has this aura of mystery developed?

First, intelligence is usually much less concrete than knowledge of the friendly situation, which Marines are likely to know with much more certainty and detail. It is commonly understood that effective intelligence is an important factor—often *the* critical factor—in mission success or failure. Isolating or measuring the specific effects of intelligence on the mission's outcome, however, is often difficult.

Second, intelligence employs specialized techniques to develop studies and products. Intelligence personnel receive certain specialized training—but hardly more than specialists in other fields. In the normal course of performing their mission, Marine intelligence sections request and receive support from specialized, technical, and sometimes highly compartmented national, theater, or service-level intelligence agencies. While these activities provide access to resources necessary to develop tactical intelligence, the activities themselves may be of limited interest to combat units. In addition, intelligence often involves highly specialized technology, especially in the collection of information.

Finally, to protect the value of a piece of intelligence as well the sources used in developing it, many intelligence products and methods are classified. Out of the legitimate concern for security, a need to know is a basic requirement for access to intelligence products. In order to protect the sensitive nature of some intelligence activities, elements of the intelligence section may be physically separated from other staff sections, with access to these elements being controlled. Unfortunately, the rightful concern over security contributes more to the mystification of intelligence than any other single factor.

The result is a veil of mystery that often surrounds intelligence activities. However, intelligence is not an obscure activity unrelated to other warfighting activities. In fact, intelligence is a central component of command and control, a fundamental responsibility of command, and inseparable from operations. All personnel involved in the conduct and support of operations—commanders, operations officers, logisticians, communicators, etc.—must understand intelligence just as intelligence personnel must comprehend the conduct and support of operations.

There is nothing mysterious about intelligence. While intelligence collection and production may involve the use of high-technology sensors and networks, good intelligence is primarily the result of solid headwork and legwork, not the output of some secret process or compartmented database. Good intelligence begins with commanders clearly identifying their intelligence concerns and needs. It is developed through the focused collection of information, thorough study, and, most importantly, effective analysis and synthesis. The result is an intelligence product that provides knowledge, reduces uncertainty, and supports effective decisionmaking.

WHAT MAKES INTELLIGENCE DIFFERENT?

We have noted that while intelligence uses specialized capabilities and techniques, this alone does not distinguish intelligence from other command and control functions. What makes intelligence unique? The one feature which distinguishes intelligence from the other command and control functions is that *intelligence deals directly with an independent, hostile will personified by the enemy.* As such, intelligence deals with more unknowns and has less control over its environment than any other aspect of command and control.

A commander may well face unknowns about the friendly situation—uncertainty about the location, activity, or status of friendly forces. Presumably such uncertainty is not the result of a conscious effort on the part of those forces to deny that information to the commander. If commanders have questions about the friendly situation, they can usually obtain the answers directly from the principals involved. In other words, for nearly every question about the friendly situation, there is a reliable source ready to provide an answer.

This is not the case for questions concerning the enemy or the area of operations. Such information by its very nature will be significantly more difficult to obtain. The enemy will do his utmost to deny us knowledge of his capabilities, dispositions, methods, and intentions through the use of security measures and his own counterintelligence. He may intentionally present us with erroneous or ambiguous information. When a foe suspects that we know something significant about his situation, he will likely undertake actions to change that situation.

This is especially true at the tactical level of war. The closer a unit is to contact with the enemy, the greater attention it pays to security, camouflage, dispersion, and deception. Moreover, once execution begins, the rapidity of changes in the tactical situation combines with the friction and fog of war to make it increasingly difficult to develop a coherent image of the enemy situation. This is why it often seems that we have better intelligence about what is happening in the enemy's rear echelon or capital city than we have about what is occurring beyond the next hill.

We have to work much harder to obtain information and knowledge about the enemy than we do concerning the friendly situation. Despite our extensive specialized capabilities designed to collect information about the enemy, the information we collect will normally be less than what we would like to have. Furthermore, collecting information does not by itself provide the needed intelligence. Even when friendly forces are obtaining information directly from the enemy—intercepting enemy signals, interrogating prisoners of war, translating captured documents—we must still confirm, evaluate, interpret, and analyze that information. Follow-on collection, processing, and production activities are normally needed. Finally, it should be emphasized that our need for intelligence usually greatly exceeds our ability to produce it; while questions about the hostile situation are almost infinite, the intelligence resources available to answer those questions are limited.

Once we have obtained the information necessary to build a picture of the enemy situation, we are confronted with other challenges. First, we must properly interpret the information. More important than the volume of information is the value of the information and the abilities of the people interpreting it. Any single item or any collection of information may be interpreted in a number of ways. Many mistakes in intelligence are not the result of a failure to collect the correct information, but rather a failure to discern the correct meaning from the information collected.

Second, even if we can develop a good understanding of the current situation, we cannot know with certainty what will happen in the future. While we can often assess the enemy's capabilities, we can rarely be certain of his intentions. Capabilities are based ultimately on factual conditions, while intentions exist only in the mind of the enemy—assuming the enemy even knows clearly what he wants to do. Thus, any assessment of enemy intentions is ultimately an estimate. While good intelligence can identify the possibilities and probabilities, there will always be an element of uncertainty in these estimates.

Third, because we are dealing directly with a hostile will, we can never be sure we are not being actively deceived. Even if we should gain some type of access to his actual plans, we cannot be certain that the enemy does not want us to see those plans as part of a deliberate deception effort.

Finally, the problems facing intelligence are further compli-
cated by the irony that good intelligence may actually invalidate itself. Consider the following instance. Intelligence estimates that the enemy is preparing to launch an attack in a
certain sector. Acting quickly on this intelligence, the commander strengthens that sector. The enemy, however, detects
our enhanced defensive preparations, which causes him to can-
cel the attack. As a result, the intelligence estimate which predicted the attack in the first place appears wrong—but only
because it was initially correct. Intelligence is thus a highly imprecise activity at best, and its effects are extremely difficult, if
not impossible, to isolate.

For example, consider the U.S. response to a movement of
Iraqi troops toward the Kuwaiti border in October 1994. After
a period of increasing tension between Iraq and the United Nations over continuing sanctions against Iraq, U.S. intelligence
detected the deployment of almost 80,000 troops in the vicinity
of the Kuwaiti border, including two elite Republican Guard
divisions.[9] The situation appeared similar to the one in August
1990 when Iraq invaded Kuwait. Intelligence warned that another invasion was possible. The United States and other allies
responded by immediately dispatching forces to the region. The
Iraqi forces were withdrawn, and the threat subsided. Did Sad-
dam Hussein intend to invade Kuwait again? We will probably
never know; intentions can seldom be determined with absolute
certainty. On the one hand, we could state that intelligence
failed because we could not as- certain Hussein's exact intentions and thus were unable to detect the difference between a

provocation and an actual invasion. A more reasonable explanation, however, is that intelligence stimulated appropriate action, action which prevented an invasion. The warning appeared to be incorrect, but only because it was right in the first place.

EXPECTATIONS OF INTELLIGENCE

We expect a great deal from intelligence. We ask intelligence to describe in detail places we have never seen, to identify customs and attitudes of societies fundamentally different from our own, to assess the capabilities of unique and unfamiliar military or paramilitary forces, and to forecast how these societies and forces will act in the future. Most notably, we want intelligence to enter the thought process of an enemy commander and predict, with certainty, what course of action he intends to pursue, possibly even before he knows himself what he is going to do. The standard against which we measure intelligence is also high. We desire a depth and degree of accuracy in our intelligence which approaches perfection. Even when a reasonable response has been provided to almost every intelligence requirement, there is still one more question to be answered, one more detail to be fleshed out, one more estimate to be refined. This is as it should be. The price for failure in intelligence is high. Inadequacies in intelligence can lead directly to loss of life, destruction of equipment and facilities, failure of a mission, or even defeat.

When properly focused and given adequate time and resources, our intelligence can come close to meeting these standards. We can provide comprehensive depictions of phys- ical terrain and manmade structures or facilities. Our reconnaissance and surveillance systems can detect and track the movements of ships, aircraft, and ground formations, in certain instances even providing real-time images of enemy activity. Our signals and human intelligence capabilities, coupled with expert analysis, can provide insight into both enemy capabilities and intentions.

However, even in the best of circumstances, intelligence still operates in an environment characterized by uncertainty. Uncertainty is a fundamental attribute of war. As discussed in the previous section, intelligence deals directly with the independent, hostile will of the enemy. This makes intelligence more susceptible to uncertainty than any other command and control function. In practical terms, this means that there are very definite limits to what commanders can reasonably expect from intelligence. Not only will more gaps exist in what we know about the enemy than in what we know about our own situation, but the reliability of everything we do know will be subject to greater scrutiny and doubt. Even if we obtain the correct information, there is no guarantee that we will interpret it correctly or that it will not change. We may be the victims of deception, whether it is by a deliberate enemy effort or by our own preconceptions. Intelligence produces estimates rather than certainties; it is important to remember that "estimating is what you do when you do not know."[10]

Intelligence may be incorrect sometimes and incomplete at other times, and it often lacks the desired degree of detail and reliability. Some of the questions asked are simply beyond knowing—or are beyond knowing given the time and resources available. Gaps in our knowledge of the enemy sit-uation, sometimes sizable, are a natural and unavoidable characteristic of fighting an enemy having an independent, hostile will. We must continually remember that *intelligence can reduce but never eliminate the uncertainty that is an inherent feature of war.*

INTELLIGENCE IN THE INFORMATION AGE

As a result of the ongoing information revolution, more people have access to more information more quickly than ever before. Intelligence has benefited greatly from improvements in information gathering, processing, and dissemination. Sophisticated sensors clandestinely collect vast quantities of data in all regions of the world. Integrated databases allow us to store and rapidly retrieve virtually unlimited numbers of reports, images, and studies. Information processors assist us in analyzing the data and developing tailored, graphic-enhanced products that convey intelligence in a more meaningful form. Communications systems give us the ability to share databases, exchange intelligence, and disseminate products almost instantaneously on a worldwide basis.

While it is alluring to believe that the information revolution will solve the problems of uncertainty in dealing with the enemy, technology has its shortcomings as well. Systems employed in intelligence can be expensive and complex. Many are controlled at the national or theater levels, where priorities might not be consistent with those of the tactical commanders. Despite their sophistication, these systems are still subject to failure as a result of weather conditions, breakdowns, or enemy countermeasures.

Further, these systems generally provide and manipulate data and information rather than generate knowledge or understanding. The information revolution has created the very real danger of information overload—more available information than can be readily used or understood. Humans have a limited capacity to assimilate information. Even if we are able to collect vast amounts of information, *information alone does not equate to knowledge or understanding*, which are ultimately the product of human cognition and judgment. Since very few pieces of information are decisive by themselves, they must be interpreted and given meaning.

Finally, the seemingly unlimited availability of information does not necessarily help us in determining which information we should collect and develop into intelligence. In an unstable international environment, in which unanticipated crises proliferate, it is difficult to identify the next enemy or potential enemy. This complicates commanders' problems of identifying their concerns and priorities; it may be harder than in the past to focus the intelligence effort. As a result, it may not be possible to develop adequate basic intelligence about potential enemies or regions well in advance.

We must continue to pursue advances offered by technology to enhance our intelligence capabilities. At the same time, we recognize that technology by itself does not produce effective intelligence. Improvements in data collection, information processing, and dissemination are tools which assist in the intelligence effort. These tools increase our capabilities only when they are applied by knowledgeable and skilled Marines focused on producing timely, useful, and relevant intelligence.

A CASE STUDY: DESERT STORM 1990–1991

The development and use of intelligence in support of Marine operations during Desert Storm illustrate the nature of intel- ligence and its core concepts and challenges. During this operation, intelligence provided an accurate picture of the situation confronting Marine forces and identified the enemy's critical vulnerabilities which Marine commanders exploited to achieve success.[11]

By mid-January 1991 the situation at the strategic and operational levels was well understood. Iraqi commanders prepared for the expected Coalition assault into Kuwait in a manner that reflected the success of their defensive strategy during the Iran-Iraq War. They constructed two major defensive belts in addition to extensive obstacles and fortifications along the coast.

Intelligence identified three Iraqi centers of gravity at the operational level. The first was their command and control. If rendered unable to direct its military forces, Iraq would not be able to mount an effective defense at the operational level. Second was Iraq's weapons of mass destruction. Degrading this capability would reduce a major aspect of the Iraqi strategic threat to other states in the region. The third center of gravity was the Republican Guard. Destroying or severely degrading the Republican Guard's ability to fight would dramatically diminish Iraq's capability to conduct a coordinated defense or to pose an offensive threat to the region later.

Intelligence likewise provided thorough understanding of Iraq's critical vulnerabilities: a rigid, top-down command and control system, the reluctance of Iraqi commanders to exercise initiative, an overly defensive approach to battle with limited ability to conduct deep offensive actions, vulnerability to air attack, an overextended logistics system, and extremely limited intelligence capabilities.

This understanding was used to plan the campaign and guide the conduct of air operations during the first few weeks of Desert Storm. Coalition air attacks had devastating effects on the Iraqis, severely disrupting their command and control, eliminating their naval and air forces, and degrading their logistics capabilities.

Nevertheless, the situation facing Marine commanders was less clear. Much of the intelligence developed prior to the start of the operation was focused on strategic- and opera- tional-level objectives and lacked the detailed, tailored intelligence essential for tactical planning. Further, ground force commanders were not permitted to employ most of their organic collection assets within Kuwait due to concerns about potential casualties, operational security, and initiation of engagements before a decision on ground operations had been made. Marines required support from national and theater intelligence agencies to answer many of their critical intelligence requirements. Although these needs were generally recognized as valid by the higher echelons which controlled these assets, Marine tactical requirements tended to fall too low on the priorities list to compete effectively with other requirements. When national sensors were used to support Marine force requirements, the results often did not provide sufficient detail to fully satisfy those requirements.

Still, by January the Marine forces' intelligence estimate provided a fairly accurate assessment of the overall size and disposition of the Iraqi units as well as the strategy and tactics they would employ. The estimate highlighted four potential Iraqi responses to Coalition air attacks: terrorist attacks inside and outside the area of operations, air and naval counterstrikes, surface-to-surface missile and multiple rocket launch- er attacks against Marine positions in the forward area, and a limited-objective ground attack or raid. Like most estimates, this assessment proved to be only partially correct. Iraqi air and naval forces offered minimal opposition, and no major terrorist attacks were conducted. However, there were extensive missile and multiple rocket launcher attacks, and a significant ground attack was launched at the end of the month.

The main shortfall of the estimate was that it lacked details required for tactical planning. Determining the effectiveness of Iraqi forces was a critical requirement. The raw numbers indicated that large Iraqi forces remained within Kuwait. Air attacks were damaging the enemy's forward echelon and had severely degraded both his sustainment and command and control capabilities. There were indications that Iraqi front-line infantry troops were demoralized and would not put up much of a fight. Intelligence clearly showed that the Iraqis had been hurt, but in the absence of definitive information detailing how badly they had been hurt, Marine planning continued to reflect a cautious approach.

During a battle from 29 January to 1 February at Al-Khafji, an Iraqi division-sized ground attack was soundly defeated by Coalition forces. Iraqi actions during this battle provided Marine intelligence specialists critical information to fill in the intelligence picture. Analysis confirmed previous assessments of the deteriorating condition of enemy units and the Iraqis' limited capability to coordinate between tactical echelons. From reports by Marine participants, analysts concluded that Iraqi soldiers were unmotivated, poorly trained, and unable to conduct combined arms operations.

With this insight, previous perceptions of the enemy's strengths and ability to mount a formidable defense were called into question. Subsequent intelligence operations further clarified the threat picture. Interrogations of enemy troops lured to surrender reinforced the view that the Iraqi will to fight was far weaker than anyone had anticipated. Artillery raids failed to elicit counterbattery fire, indicating that Iraqi artillery capabilities had been degraded. Unit boundaries along the defensive belts were located, and gaps in the defenses were identified at those points, confirming that coordination between tactical echelons was poor.

From the new intelligence, a new estimate reflected the likelihood that the Iraqis would be unable to conduct an effective defense of the forward positions. It noted that the Iraqis could not coordinate between units, employ supporting arms, or conduct counterattacks with forces larger than a battalion. Finally, it indicated that Iraqi infantry and artillery troops would probably surrender en masse once the first shot was fired.

This intelligence was used to substantially revise the Marine operation plan. Knowing that the Iraqis would be unable to assess what was happening on the battlefield or to respond effectively, Marine commanders adopted a more aggressive scheme of maneuver. The previous plan called for a sequential attack with one division following the other through the defensive belts using a single breech point. On February 6, the Marine Force commander approved a new plan in which two Marine divisions would conduct a simultaneous, coordinated attack through the defensive belts at points 40 kilometers apart (see figure 1). In addition, a significant force was now assigned to deal with the expected flood of surrendering Iraqi troops.

The intelligence assessment developed and refined during Operation Desert Storm reduced uncertainty, enhanced situational awareness, and aided Marine commanders in planning and decisionmaking. This assessment did not answer every question, but it did identify the enemy's critical vulnerabilities which were exploited to achieve decisive results.

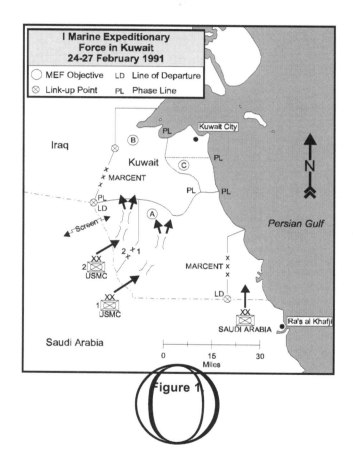

Figure 1

CONCLUSION

Intelligence is a fundamental component of command and control—inseparable from both command and operations. Accurate, timely, and relevant intelligence is critical to the planning and conduct of successful operations. Effective intelligence uncovers enemy weaknesses which can be exploited to provide a decisive advantage. Shortfalls in intelligence can lead to confusion, indecision, unnecessary loss of life, mission failure, or even defeat.

Intelligence is knowledge of the enemy and the surrounding environment provided to support the commander's decisionmaking process. Intelligence is more than just information; it is the analysis and synthesis of information which provides a meaningful assessment of the situation. Intelligence evaluates existing conditions and enemy capabilities, estimates possible future conditions and enemy courses of action, assists in the development and evaluation of friendly courses of action, and aids in protecting friendly forces against the effects of enemy action.

While intelligence uses specialized capabilities and techniques in developing a useful product, it is not an obscure process isolated from other warfighting functions. In fact, effective intelligence requires a firm focus on the needs of commanders. This in turn, demands integration with all aspects of operational planning and execution. What separates intelligence from the other aspects of command and control, however, is the fact that intelligence must deal directly with the independent, hostile will of the enemy. Because intelligence attempts to look into the future despite significant unknowns, the resulting product will always be based on estimates, not certainties. Users of intelligence must always remember that intelligence can reduce, but never eliminate the uncertainty that is an inherent characteristic of war, and act accordingly.

Chapter 2

Intelligence Theory

"Many intelligence reports in war are contradictory; even more are false, and most are uncertain. What one can reasonably ask of an officer is that he should possess a standard of judgment, which he can gain only from knowledge of men and affairs and from common sense. He should be guided by the laws of probability. These are difficult enough to apply when plans are drafted in an office, far from the sphere of action; the task becomes infinitely harder in the thick of fighting itself, with reports streaming in."[1]

—Clausewitz, *On War*

"It will be vital to identify centers of gravity rapidly and determine the critical vulnerabilities that will be our pathways to them. We won't always have the luxury of a passive foe, and there's no natural law that says that every high-tech war must be fought in a desert with unlimited visibility and good weather."[2]

—Carl E. Mundy

Having reached a common understanding of the nature of intelligence, we turn now to developing a theory about the intelligence process that in turn will serve as the basis for creating an effective intelligence system.

WHAT DO WE WANT TO KNOW ABOUT THE ENEMY?

In war, it is easier to defeat an enemy we understand, even partially, than to fight an enemy who is an enigma. Intelligence is the means by which we come to understand the enemy. What is it about the enemy that commanders need to learn? The commander needs to gain knowledge at a variety of levels, ranging from that which is quantifiable to that which is purely intangible.

Obviously, we want to know the measurable things: the number of enemy personnel, armored vehicles, artillery pieces, and aircraft. We hope to learn the dispositions, organization, and locations of enemy forces. We also want to obtain technical specifications: the performance characteristics of enemy tanks and aircraft and the range and effectiveness of opposing weapons. Traditionally, intelligence has focused on these tangible factors. They usually provide a concrete image of the threat and the nature of its combat power. These tangible factors and images thus provide the foundation for de- veloping a more complete understanding of the enemy based upon other intangible factors.

While numbers definitely matter, they provide only partial insight into enemy capabilities. Less quantifiable and more subjectively deduced is the enemy's level of readiness—the state of his training, the quality of his leadership, and the morale of his forces. Commanders need to know the enemy's methods—his doctrine, tactics, techniques, and procedures—as well as his past performance in training and in combat. Even this is not enough. The successful commander cannot truly know an enemy simply through analysis of his physical environment, material might, and political and military institutions.

We must seek still more deeply for those moral and cultural forces which shape the enemy's actions. War is ultimately a human conflict, and much of human nature is decidedly illogical and incalculable. Developing sound intelligence requires an understanding of the institutions, preferences, and habits of a different culture. Commanders must appreciate the values, goals, and past experiences which motivate the enemy. We must gain insight into *why* he fights. To know what motivates an enemy to action requires an identification and appreciation of what the enemy holds dear.

Coming to grips with the intangible aspects of the enemy situation is much more difficult than assessing those factors that are quantifiable, but it is essential if commanders are to truly understand an enemy. This is exactly what intelligence must endeavor to do—to understand what factors shape an enemy's behavior in order to describe or explain that behavior.

When we try to understand an enemy, it is important to visualize the enemy as he sees himself and to see the situation as he views it. While gaining an objective appreciation for the enemy's capabilities is important, it is equally important to appreciate how the enemy perceives his own capabilities, since it is this image that will have the greatest influence on his actions. The enemy will do what *he* thinks is possible, not what we think he can do.

Seeing the situation from the opponent's perspective is especially important when confronting an adversary with a significantly different set of cultural or societal values. Enemy behavior which appears irrational—and therefore surprising—to us may in fact reflect perfectly reasonable and even predictable actions, given the enemy's values or the information available to him at the time. Consider the use of the banzai charge and kamikaze attacks by the Japanese in World War II or employment of women and children as shields to cover the actions of Somalian clansmen against U.N. forces. None of these actions is rational when viewed from a Western perspective. However, within the context of Japanese or Somalian societal norms, they are less surprising. A commander who fails to understand the enemy on his own terms risks a fundamental failure to understand the very nature of the conflict.

Finally, intelligence should strive to determine not only the enemy's capabilities but also his intentions. Intelligence estimates deal in both: they describe what an adversary *can* do, and they attempt to discern what he *will* do based upon possibilities and probabilities. Ideally, intelligence should identify several possible scenarios, answering questions such as: *Which is the most likely enemy course of action? Which is the most dangerous?* Intelligence should also correlate the effect of possible enemy actions on friendly plans.

The goal is, therefore, to obtain as complete an understanding of the enemy as possible. The ultimate purpose in seeking this understanding is to *identify enemy centers of gravity and critical vulnerabilities and other limitations that may be exploited* to defeat him: weaknesses in specific warfighting capabilities, numbers, or dispositions; technical deficiencies in weapon systems; or shortcomings in readiness, leadership, or morale. Conversely, intelligence should provide warning of particularly dangerous enemy capabilities that pose a threat to friendly forces.

There are practical limitations to achieving this level of understanding. The effects of uncertainty on intelligence have already been discussed, but it is worth repeating that gaps in our knowledge of the enemy situation are natural and unavoidable. Furthermore, attaining in-depth knowledge on the variety of potential threats confronting Marine forces is a significant challenge. Our education, cultural appreciation, and experience may be insufficient to provide detailed insight into an adversary without extensive study. The time required for this study may not be available when responding to a developing crisis. In these situations, we build as complete a picture of the enemy as possible in the time available—relying primarily on known factors and likely capabilities—while striving to fill in that picture rapidly as our understanding of the adversary grows.

An example of this approach can be seen in the history of Marine forces' involvement in Somalia. When the 15th Marine Expeditionary Unit landed in Mogadishu to begin Operation Restore Hope in December 1992, Marines had only a basic understanding of Somalian culture, the clan structure, and the threat it presented to U.S. and U.N. forces. Intensive intelligence development, in particular through human intelligence operations, rapidly increased our level of understanding. This intelligence was used over the course of the campaign to plan and execute numerous successful tactical actions that in turn further enhanced the intelligence picture. When the Marines of I Marine Expeditionary Force (MEF) returned in February of 1995 to assist in the final withdrawal of U.N. forces, they were able to draw on an extensive reservoir of intelligence to plan and execute Operation United Shield. Based on this intelligence, I MEF prepared a "playbook" of responses to cover virtually any Somalian reaction. Further, the playbook evolved as new intelligence was produced and the intelligence estimate changed. The detailed intelligence available contributed directly to the safe and effective accomplishment of the mission.[3]

CHARACTERISTICS OF GOOD INTELLIGENCE

In the previous chapter, we discussed our expectations of intelligence—its capabilities and limitations as well as what can and cannot be reasonably expected of intelligence. As the next step in developing a theory of intelligence, it is important to describe the characteristics of good intelligence.

First, intelligence should be *objective*—as free as humanly possible of bias or distortion. We have already noted that a significant problem in intelligence is not the lack of information, but the difficulty in interpreting that information correctly. Intelligence can be distorted if we attempt to make it conform to preconceived notions, fail to view the situation from the enemy's perspective, or manipulate the intelligence product to support a particular decision or conclusion. For example, prior to Chinese intervention in the Korean War, there were ample indications and warnings of imminent Chinese involvement. However, despite availability of much factual information to the contrary, strongly held preconceptions led commanders and their intelligence officers to conclude that the Chinese would not intervene.[4] Unfortunately, data and information are almost always susceptible to more than one interpretation and can be manipulated consciously or uncon- sciously to support preconceived notions.

Second, intelligence should be *thorough*, meaning that it satisfies the intelligence requirements of the commander. Thoroughness does not imply completeness and certainty to the last detail, but rather sufficient depth to assist the commander in reaching sound decisions and developing effective plans. Intelligence personnel should not only identify for the commander what is known but also what is not known. The commander may then assess the risks and decide what actions are worth these risks.

Third, intelligence should be *accurate*, meaning that it should be factually correct. Sound estimates of the enemy's capabilities and intentions must agree with the facts at hand. Since intelligence cannot be precise to the last detail, commanders must have an appreciation of the reliability of a particular intelligence assessment or product.

Fourth, intelligence must be *timely*, meaning that it must arrive in the hands of appropriate decisionmakers in time to affect tactical decisions. Intelligence does not exist for its own sake, but as the basis for taking effective action. The most accurate and valuable piece of intelligence is of no use if it arrives too late to be acted upon. Some kinds of intelli- gence are more time-sensitive, or perishable, than others—a warning report, for example, is a type of intelligence product that tends to be highly perishable. It is important to keep this time-sensitivity in mind when dealing with any intelligence product.

Fifth, intelligence should be *usable*, appearing in a form meaningful to and easily assimilated by decisionmakers. Good intelligence should be concise and clear. It must create coherent images—meaningful mental pictures that are immediately and easily understood—rather than present the commander with a mass of unfocused data. Because we generally understand information better when it is presented in the form of images, we attempt to present intelligence in a visual format whenever possible.

Sixth, intelligence should be *relevant* in that it supports the commander's planning and decisionmaking requirements. Relevance means that intelligence is pertinent to the level of command for which it is intended. Relevance means also that commanders are provided information and intelligence bearing significantly on the situation at hand and that they are not burdened with information and intelligence of minimal or no importance. Intelligence that is tailored appropriately for one commander may be too generic or too detailed for commanders above and below that particular level. However, it may be extremely difficult to know in advance what is relevant and what is not. This leads again to the necessity for commanders to focus the intelligence effort.

The value of providing relevant intelligence is illustrated by the following example. Following the bombing of the Marine Corps compound in Beirut in 1983, Department of Defense investigators faulted commanders and intelligence for inundating on-scene commanders with information and failing to provide them with timely intelligence tailored to their specific operational needs.[5] While information overload was certainly not the only cause of the Beirut tragedy, more focused intelligence might have helped commanders prevent its occurrence or at least take greater security precautions.

Finally, intelligence must be *available*—which means that it is readily accessible to appropriate commanders. Availability is a function of both timeliness and usability, but it is also a function of an effective information management system that allows commanders at various levels to readily access the intelligence they need. Availability also means that relevant basic intelligence has been developed in advance and that intelligence assets are maintained in readiness to develop other intelligence products as needed. Finally, availability is a function of effective use of security classifications that protect sources of information while at the same time ensuring that commanders have reasonable access to intelligence.

This discussion is not meant to specify a checklist for what does or does not constitute good intelligence, but to describe the general characteristics which effective intelligence tends to exhibit to one degree or another. Few intelligence products will exhibit all the above characteristics. Some of the characteristics such as timeliness, usability, and availability are mutually supportable. Others such as timeliness and thor- oughness can be in conflict. The extent to which actual intelligence demonstrates each of these characteristics depends on the particular situation.

CLASSES OF INTELLIGENCE

If we could describe a complete intelligence picture—one that provides us everything we need to know about a given situation—that description would include knowledge of established conditions which have existed in the past, unfolding conditions as they exist in the present, and conditions which may exist in the future. Our complete image would include what has been, what is, and what might be. With this background, two classes of intelligence are defined. The first is *descriptive intelligence,* which describes existing and previously existing conditions. The second class, which attempts to anticipate future possibilities and probabilities, is *estimative intelligence.*

Descriptive intelligence has two components. The first is *basic intelligence.* Basic intelligence is general background knowledge about established and relatively constant conditions. Basic intelligence is often compiled in advance of potential operations and retained in databases or reference publications. Basic intelligence might describe the geography, culture, economy, and government institutions of a potentially hostile nation or area. With regard to the military capa- bilities and limitations of potential enemies, basic intelligence might detail the size, organization, and equipment of their military forces. These factors may change, but only slowly.

Basic intelligence is often encyclopedic in nature and is consequently often the most mundane. While it tends to be the easiest to gather, often being available through open sources, the depth and detail of the intelligence required to support most operations makes developing basic intelligence a labor-intensive and time-consuming task. Of all the types of intelligence, basic intelligence tends to be the most accurate and reliable. However, basic intelligence is also the most general and least time-sensitive. By itself it rarely reveals much that is decisive. Further, since basic intelligence does not address specific situations, it rarely provides sufficient knowledge for effective decisionmaking. Nevertheless, basic intelligence establishes the necessary foundation for building a more complete intelligence picture.

Descriptive intelligence also includes *current intelligence,* which is concerned with describing the existing situation. In general, current intelligence describes more changeable factors than those addressed by basic intelligence and is therefore more time-sensitive than basic intelligence. For example, while basic intelligence reports climatic norms, current intelligence describes existing weather conditions and its effects on operations; while basic intelligence shows enemy doctrine and

43

organization, current intelligence depicts actual dispositions, movements, and patterns of activity. At higher levels, basic intelligence describes economies and forms of government; current intelligence addresses ongoing enemy war preparations or the status of relations with other hostile or potentially hostile nations. As a rule, current intelligence tends to be more specific than basic intelligence but less reliable and harder to obtain. Basic intelligence provides the broad picture upon which current intelligence expands by adding specific details about the existing situation.

Estimative intelligence, the second class of intelligence, focuses on potential developments. Developing estimative intelligence is perhaps the most important and at the same time most demanding task of intelligence. Estimative intelligence evaluates the past as delineated by basic intelligence and the present as described by current intelligence and seeks to anticipate a possible future—or several possible futures. It is concerned with determining when, where, how, or even if an enemy or potential enemy will attack or defend. Commanders cannot reasonably expect estimative intelligence to precisely predict the future; rather, estimative intelligence deals with the realm of possibilities and probabilities. It is inherently the less reliable of the classes of intelligence because it is not based on what actually is or has been, but rather on what *might* occur.

Although described as conceptually distinct, the two classes of intelligence are inseparable. Descriptive intelligence provides the base from which estimative intelligence assesses possible or probable futures. Without both classes of intelligence, it is impossible to develop a full image of a hostile situation.

CAPABILITIES VERSUS INTENTIONS

Another way to frame this same discussion is in terms of capabilities and intentions. Descriptive intelligence attempts to discern enemy capabilities and existing conditions. It attempts to answer the questions: "What conditions currently exist? What *can* the enemy do? What *can't* he do?" Estimative intelligence attempts to discern enemy intentions and future conditions. It asks: "What conditions will probably exist in the future? What are their effects on friendly and enemy capabilities and courses of action? What *might* the enemy do? What is the enemy *most likely* to do?" Although the answers to all these questions are estimates rather than certainties, generally we can assess enemy capabilities with greater precision than enemy intentions.

Estimating enemy capabilities is largely a matter of interpreting the facts. Estimating enemy intentions, however, is a matter of far less certainty. Intentions exist only in the enemy's mind. In any given situation, an enemy commander will probably have several courses of action available. There may be little or no indication of which one he favors. He may be intentionally attempting to conceal his intentions from us, or he may be trying to keep more than one option open. He may be gripped by indecision and not know what he intends to do. Furthermore, he can change his mind. He can respond to changes in the situation or our own actions in ways we cannot anticipate.

Complicating this problem is the reality that an enemy's intentions are normally the product of thought processes different from our own. We are sometimes surprised when an enemy takes an action which we consider to be irrational. However, when viewed from the perspective of the enemy's cultural norms or values, his actions may be perfectly logical and predictable. Unfortunately, it is extremely difficult to gain the depth of insight needed to understand the thought process of each and every potential adversary we face. Our own values and cultural background will always be a significant obstacle in estimating the intentions of terrorists willing to blow themselves up in suicide bombings or a dictator who would inflict massive damage on the environment by setting hundreds of oil wells on fire.

Enemy capabilities and enemy intentions are closely related. Capabilities establish the limits of intentions; the enemy cannot intend to do something beyond his capabilities and accomplish it successfully. However, it is crucial to note that it is not actual capabilities that matter, but the enemy's perception of his capabilities. The enemy will act based on his perception of his capabilities. His perception may or may not agree with our evaluation of what he can or cannot do. Our analysis of enemy forces in South Vietnam did not credit the North Vietnamese and Viet Cong with the capability to launch a nationwide offensive in 1968. The Tet offensive of that year clearly demonstrated that the enemy believed other- wise.[6] Again, a key element in assessing both capabilities and intentions is the ability to view the situation as the enemy perceives it.

Analysis of capabilities and analysis of intentions are by no means incompatible. Any effective intelligence picture must provide insight into both. Without some appreciation of enemy intentions, it is extremely difficult to decide on an effective plan of action. However, without an understanding of the enemy's capabilities, it is impossible to estimate his intentions.

SIGNALS AND NOISE

Complicating our ability to assess capabilities and estimate intentions is the problem of interpretation of the information we collect. To develop objective and accurate intelligence, we must understand this problem. We can examine it through a discussion of *signals* and *noise*.[7]

Signals refer to those pieces of information commanders re- ceive that, if properly interpreted, can lead to valuable in- sight about the situation. Signals help with our situation assessment. Noise, on the other hand, refers to various pieces of useless in- formation—information which is false, out of date, inaccurate, ambiguous, misleading, or irrelevant. An enemy may intention- ally present a foe with noise in order to mislead, but noise is not necessarily the product of an enemy deception. Like static on a radio, noise interferes with our reception and interpreta- tion of valuable signals.

The difficulty is to distinguish signals from noise. Unlike the distinction between radio static and the true signal, the differ- ence between true and false information is rarely easy to distin- guish. We endeavor to identify critical enemy vulnerabilities, but signals of such vulnerabilities are rarely clear at the time. The recognition of what is important, relevant, and accurate sometimes becomes clear only in hindsight—if then.

Clausewitz advises that, "A sensitive and discriminating judgment is called for; a skilled intelligence to scent out the truth."[8] Just as judgment is no substitute for good intelligence, *intelligence is no substitute for good judgment.* The two must go hand in hand.

This point illustrates the importance of mindset to intelli- gence. A mindset is a set of assumptions, biases, and precon- ceptions. A mindset reflects a preexisting image of what is reasonable; it serves as a filter that helps to distinguish the signals from the noise. The human tendency is to be more receptive to information that is consistent with one's mindset and more skeptical of information that is not. In other words, information that is consistent with an existing mindset is interpreted as signals, and information that is not is construed as noise.[9]

Every individual possesses a unique mindset. Biases and preconceptions are also indispensable to intelligence. Without them, it would be impossible to make sense of the available mass of confusing and sometimes contradictory information. Mindsets serve as a frame of reference, enabling us to quickly categorize and assess the relevance and reliability of vast amounts of information. Without a preexisting mindset, commanders would likely be overwhelmed by the amount of information and unable to distinguish between signals and noise.

At the same time, mindsets always bring the danger that we will subconsciously interpret intelligence to comply with our preconceived notions rather than with reality. This hazard applies equally to the producers and users of intelligence.

The danger of preconceived notions and their impact on signals and noise is illustrated by this example from the Arab-Israeli War of 1973.[10] In October 1973, a simultaneous Syrian and Egyptian attack caught the Israeli army badly unprepared. There had been plenty of signals of the Egyptian intentions.

The Israeli high command had extremely detailed and accurate information on their enemies' order of battle, unit locations, armaments, and readiness status. In fact, Israeli intelligence was fully aware of unprecedented forward de- ployments of enemy troops and ammunition stocks. Nonetheless, the possibility of war was discounted until just 8 hours prior to the beginning of the Arab offensive.

The problem was not lack of information, but an inability to filter out noise coupled with reliance on a set of rigid and faulty preconceptions. The Egyptians employed a variety of deception activities, both political and military, to create noise. Included were the continued preparation of defensive positions, repeated repositioning of units along the front, and the use of training exercises as a cover for forward deployments.

Despite the effort to deceive, Israeli intelligence detected most of the key preparations for war. However, even though there was information pointing to an impending attack, the Israeli mindset prevented accurate interpretation of it. The Israelis believed that any Arab attack would be based on military rather than political or psychological factors. Since the Israelis felt that Syrian or Egyptian armies had no capability to conquer substantial territory from Israel, they discounted any possibility that the Arabs would attack. In fact, the Arab objective was not territorial conquest, but creation of a diplomatic crisis that would be resolved in their favor. Compounding the Israelis' misreading of Arab intentions was the failure of Israeli intelligence to objectively assess Egyptian military capabilities.

The Israelis' preconception that the Egyptian army was incompetent caused them to dismiss any possibility that the Egyptians could mount an effective attack. Due to Israeli bias and preconceptions, an overwhelming body of first-rate intelligence that would have provided adequate warning of the attack was simply dismissed as irrelevant.

We seek to develop a balanced mindset that provides a sensitive and discriminating judgment which is not so entrenched that it deafens us to alternative signals. One of the most valuable contributions intelligence personnel can provide is unbiased analyses to uncover and guard against dangerous preconceptions.

LEVELS OF INTELLIGENCE

A complete intelligence picture must also provide insight into the enemy as a complete entity or system, not merely as a collection of unrelated individuals, units, or organizations. Companies, batteries, or squadrons normally do not act independently. They conduct operations in accordance with the plans and orders of a senior headquarters that in turn is attempting to achieve some strategic or operational objective. In order to understand what the enemy unit directly opposite us is doing now or what it might do in the future, it is usually necessary to examine the capabilities and intentions of enemy units

51

and commanders two levels or more above our immediate adversary.

Developing this type of understanding requires us to consider that intelligence cuts across the three levels of war: tactical, operational, and strategic. As this intelligence varies in terms of scope, application, and level of detail, we divide intelligence into levels which correspond to the levels of war. *Tactical intelligence* concerns itself primarily with the location, capabilities, and possible intentions of enemy units on the battlefield and with the tactical aspects of terrain and weather. *Operational intelligence* pertains more broadly to the location, capabilities, and possible intentions of enemy forces within the theater and with the operational aspects of geography. Finally, *strategic intelligence* is broadest of all in scope and addresses the factors needed to formulate policy and military plans at the national and international levels.

Marine Corps intelligence focuses on tactical intelligence, which is the level of intelligence Marines need, generate, and use most often. However, in order to operate effectively, Marine forces require ready access to operational and strategic intelligence, as well as tactical, to comprehend the larger situation and provide appropriate context for the development of tactical intelligence products.

INTELLIGENCE REQUIREMENTS

A unit's intelligence effort begins with receipt of the mission and the commander's guidance. On-hand intelligence is rarely sufficient to support comprehensive planning and decisionmaking needs—gaps will remain. Such intelligence gaps are known as *intelligence requirements*.

Intelligence requirements are questions about the enemy and the environment, the answers to which a commander requires to make sound decisions. The breadth of potential intelligence gaps, however, will generally far exceed organic intelligence capabilities. Thus, it is important to focus intelligence operations on those intelligence requirements crucial to mission success. We call these requirements *priority intelligence requirements*.[11]

Priority intelligence requirements are intelligence requirements associated with a decision that will critically affect the overall success of the command's mission. Priority intelligence requirements constitute the commander's guidance for the intelligence collection, production, and dissemination efforts.

The nature and scope of intelligence requirements will vary with the level of command and its mission. Further, the type of operation and at what particular phase of planning or execution the commander states a requirement will be major influences on its breadth and complexity. However, it is the commander who designates the priority intelligence requirements and there-fore provides direction to the intelligence effort.

The importance of clear and focused intelligence require-ments was demonstrated during the recovery of Air Force pilot Capt Scott O'Grady from Bosnia in June 1995. After Capt O'Grady was shot down, the 24th Marine Expeditionary Unit (Special Operations Capable) afloat in the Adriatic was placed on alert to conduct a tactical recovery of aircraft and personnel (TRAP) mission. Upon notification that a signal had been re-ceived from Capt O'Grady's recovery beacon, the MEU com-mander designated three priority intelligence requirements: determine Capt O'Grady's exact location, update and reassess air defense threats to the TRAP force, and identify and de-scribe flight hazards to the TRAP force. The clear identifica-tion of the critical intelligence needed to execute the mission enabled the MEU intelligence section to concentrate its effort on satisfying those requirements in the few hours it had available.[12]

SOURCES OF INTELLIGENCE

Intelligence information comes from a wide variety of informa-tion sources, ranging from a reconnaissance Marine with a pair of binoculars to a sophisticated electronic sensor. Intel- ligence information may be derived from any Marine, other human sources, imagery, radar, signals, other emissions or signatures, and open-source references. Effective intelligence operations employ all information sources, whether organic or external. The value of a source is not necessarily related to the sophistication or cost of that source. Sources of information must be appropriate to the nature of the particular intelligence requirement; that is, the collection method or capability used must be appropriate to the aspect of the enemy or the en- vironment about which information is needed. For example, electronic intelligence will likely be of little use against a technologically unsophisticated enemy; human intelligence sources will generally be more valuable. We must tailor the sources to the requirement, ensuring that we exploit both the observations of units in direct contact with the enemy and our more sophisticated sensors.

FUNCTIONS OF INTELLIGENCE

Intelligence performs six specific functions in order to provide knowledge of the threat and the surrounding environment as well as to deny that same knowledge to the enemy.

The first function of intelligence is to *support the formulation of the commander's estimate of the situation* by providing as accurate an image of the hostile situation as possible. Through this function, intelligence helps the commander gain an initial appreciation for the terrain, weather, and other aspects of the operational environment. Intelligence personnel use techniques (such as intelligence preparation of the battlespace) to estimate enemy capabilities, intentions, vulnerabilities, and possible courses of action. In this manner, intelligence supports initial decisionmaking and planning.

The second function of intelligence is to *aid in situation development*—to provide continuing knowledge of unfolding events to help update the commander's image of the hostile situation. While the commander's initial estimate of the situation takes place before execution and provides the basis for the plan, situation development occurs during execution and provides the basis for adjusting plans to adapt to new circumstances or to exploit opportunities as they arise.

The third function of intelligence is to *provide indications and warnings*. Indications and warnings serve a protective purpose, namely to provide early warning of potential hostile action and thereby lessen the chance of being surprised. Properly used, indications and warnings act as alarms. They alert us to developments that run counter to the commander's planning assumptions and understanding of the situation in time to take necessary actions or precautions.

A fourth function of intelligence is to *provide support to force protection*. Force protection includes defensive operations, security measures, and collection activities undertaken by a commander to guard the force against the effects of enemy action. Intelligence supports the commander's force protection needs by estimating an enemy's intelligence, ter- rorism, espionage, sabotage, and subversion capabilities as well as recommending countermeasures against those capabilities. Support to force protection requires detailed assessments of both the capabilities and intentions of the enemy. A successful program of force protection lessens the enemy's ability to take offensive action against us.

The fifth intelligence function is to *support targeting,* a function that intelligence shares with operations. Targeting is the process of acquiring information about targets and choosing the best method for attacking those targets. Intelligence supports this process by locating and portraying targets for attack and by estimating the vulnerability and relative importance of those targets. Targets may be physical targets such as a bridge or enemy position, or they may be functional targets such as the enemy's command and control system.

The final role of intelligence is to *support combat assessment.* Combat assessment is the process used to determine the effects of friendly actions on the enemy. It includes battle damage assessment which refers specifically to the effects of friendly fires on enemy targets. It also applies more broadly the overall effects of friendly actions on enemy capabilities and intentions. Combat assessment provides the basis for future friendly actions as well as a dynamic link back to the first step of the intelligence cycle.

SECURITY

Safeguarding intelligence is an essential consideration. Intelligence is normally less valuable if the enemy is aware of what we know. If the enemy concludes that we are in posses- sion of a key piece of intelligence, he will likely change his plans and thus invalidate the intelligence. Security is important not only because it protects a specific piece of intelligence but also because it protects the sources upon which the intelligence is based. Thus, in the interests of security, the dissemination of and access to intelligence is often restricted.

A tension exists between the legitimate need for security and the essential need for dissemination. On the one hand, we must protect not only the value of individual pieces of intelligence but also the sources which we depend upon to provide additional valuable intelligence in the future. On the other hand, intelligence is useless unless it can be acted upon; to be of value, intelligence must be in the hands of the decisionmakers who plan and execute military operations. Finding the proper balance between greater security and wider dissemination is a matter of reasoned judgment based upon the situation, the nature of the intelligence, and the sources involved.[13]

An example of the judgments involved in balancing security and dissemination can be seen in the use of signals in-telligence during World War II. The Allies had significant success in breaking both German and Japanese codes during the war. Because of the sensitivity of the intelligence derived from communications intercepts, the desire to ensure continued availability of this source of intelligence, and the ease with which the enemy could have taken measures to protect their communications, access and dissemination were tightly controlled. Commanders were faced with difficult choices in deciding when and how to use this intelligence, weighing the potential gain against the risk of compromising the source of intelligence. For example, when U.S. cryptologists intercepted advance notification of an inspection tour of the forward area by the commander in chief of the Japanese imperial combined fleet, Admiral Isoroku Yamamoto, American commanders had to decide whether or not to ambush Yamamoto's plane. In this case, they deemed the potential gain of elim- inating Japan's best military leader worth the risk of compromising the source of the intelligence. U.S. forces were able to shoot down Yamamoto's plane, resulting in his death—without compromising any U.S. intelligence sources.[14]

THE INTELLIGENCE CYCLE

The intelligence cycle describes the general sequence of activities involved in developing intelligence. The cycle is not meant to prescribe a procedure to be followed, but simply to describe a process which generally occurs. The intelligence cycle has six phases through which information is planned, obtained, assembled, converted into intelligence, provided to decisionmakers, and, ultimately, used in making decisions. (See figure 2.)

The first phase in the intelligence cycle is *planning and*

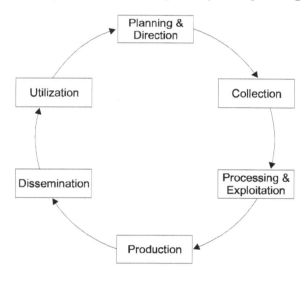

Figure 2.

direction. This phase consists of the identification of intelligence requirements and the planning of intelligence oper- ations and activities to satisfy those requirements. The commander directs the intelligence effort; the intelligence officer manages this effort for the commander. In so doing, the intelligence officer is guided by the commander's intent, the established priority intelligence requirements, and specific guidance provided by the commander for the conduct of the intelligence effort. Planning and direction encompasses the supervision of collection, processing, production, and dissemination operations as well as developing the intelligence structure necessary to support planned or ongoing operations.

Collection is the second phase of the intelligence cycle. During collection, organic, attached, and supporting intelligence sources collect and deliver information to the appropriate processing or production unit—or, in some instances, directly to the appropriate commander for immediate action. Effective collection depends upon the use of a variety of mutually reinforcing sources. Necessary, planned redundancy and overlap of sources increase the reliability of information and can reduce the effectiveness of enemy deception or denial efforts.

Processing and exploitation is the third phase of the intelligence cycle, the conversion of raw data into a form suitable for the production of intelligence. Largely a technical function, processing and exploitation converts the data into an understandable form and enhances its presentation. Examples of processing and exploitation include developing and interpreting a piece of film, translating a foreign-language text, or decoding an encrypted radio report. Not all information requires processing; some is collected in a form already suitable for production. Sometimes processing and exploitation occurs automatically during collection.

The fourth phase of the intelligence cycle is *production*, the activities by which processed data is converted into intelligence. Production involves evaluating the pertinence, reliability, and accuracy of information. It involves analyzing information to isolate significant elements. It includes integrating all relevant information to combine and compare those elements of information with other known information. Finally, production involves interpreting the information to form logical conclusions that bear on the situation and that support the commander's plan to engage the enemy. *Production is a process of synthesis—the most important action in developing usable intelligence for the commander.* Production arranges the intelligence pieces to form coherent images. It is this step which adds meaning to these pieces, creating *knowledge*. Synthesis does not generally create a complete image—totally filling in the gaps and eliminating uncertainty—but it should provide an image from which the commander can reach an acceptable level of understanding. In the end, synthesis answers the all-important question: *"What effect does all of this have on our ability to accomplish the mission?"*

The fifth phase of the intelligence cycle is *dissemination*, the timely conveyance of intelligence in an appropriate form and by a suitable means to those who need it. Depending on its importance and time-sensitivity, intelligence may be disseminated—*"pushed"*—directly to users, or it can be sent to an accessible data base from which commanders can *"pull"* that intelligence which they need (see figure 3). Intelligence flows by any number of channels or methods. The form intelligence takes can influence dissemination. Some intelligence can be transmitted almost instantaneously to multiple users via a digital communications link, while other intelligence must be physically delivered by courier. The channel or means of dissemination is less important than the arrival of the intelligence at the proper destination on time and in a form readily usable to the commander. Depending on the urgency and time-sensitivity of the intelligence, it may follow established communications channels, or it may be broadcast to the entire force simultaneously as an alert or alarm.

The final phase in the intelligence cycle is *utilization*. The commander may provide direction, information may be collected and converted into intelligence, and the intelligence may

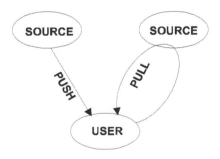

Figure 3.

be disseminated, but unless that intelligence is exploited through decision and action, it has served no purpose. Utilization is not a function of intelligence per se, but rather of command and control—making the decision and then carrying it out. This reinforces two important points made earlier: first, intelligence has no value for its own sake but assumes value only when acted upon; and secondly, intelligence is inextricably linked to command and control.

No one phase of the intelligence cycle is more important than the others—they are interdependent. Without proper direction, the other phases will be uncoordinated and ineffective. Without effective collection, there may be too much or too little information, and the information obtained may prove irrelevant. Without processing and production, the resulting mass of information may appear meaningless. Lengthening production time will delay dissemination. The first four phases of the intelligence cycle offer marginal value unless the intelligence arrives to the right person in time and in a useful form to support decisionmaking. Finally, intelligence operations are wasted if commanders fail to understand and act upon the knowledge intelligence offers. For simplicity, the intelligence cycle is described as a sequential method; however, in practice, it is a dynamic process responsive to changes in the situation and the commander's evolving in- telligence needs.

A CASE STUDY: VIETNAM 1972

The method used to produce the U.S.'s intelligence assessment of North Vietnam's intentions for 1972 provides an example of the intelligence concepts discussed in this chapter.[15] Forecasting the scope and intensity of the North Vietnamese Army (NVA) and Viet Cong (VC) operations within South Vietnam after the southwest monsoon season ended was the intelligence challenge.[16]

In seeking answers to these questions, intelligence analysts focused on a few key aspects of NVA and VC capabilities. The NVA logistics system often provided a reliable indicator of future activity. The level of NVA resupply activity usually reflected the scope and intensity of planned military operations. A related question was to determine the number of North Vietnamese soldiers moving into South Vietnam and, more specifically, to find out whether these were reinforcements or merely replacements for combat losses of the previous year.

Although there were weaknesses, by 1971 the overall quality of intelligence concerning NVA operations was good. The workings of the enemy's supply system, which had remained relatively unchanged for two decades, were well understood (see figure 4, page 66). Aerial infrared and radar imagery identified "hot-spots" of activity along the principal NVA resupply route, the Ho Chi Minh trail, and provided indications of the

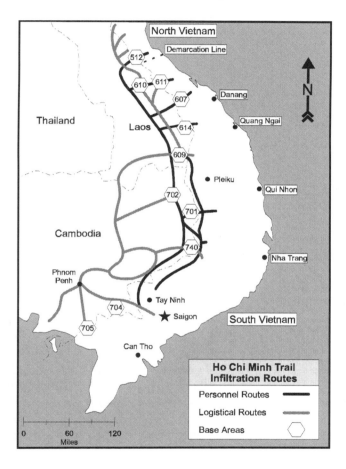

Figure 4.

intensity of that activity. Remote ground sensors placed along-side key chokepoints transmitted data on the density and type of vehicular traffic. Long-range ground reconnaissance patrols, signals intelligence, translation of captured documents, and enemy prisoner of war interrogations all helped verify the accuracy of information collected by technical means and improved the overall intelligence picture. There were some problems in the intelligence system as well, one of which was the inability of U.S. intelligence agencies to process, analyze, and synthesize the huge volume of information collected by the technical sensors and other sources.

Analysis of the NVA logistics system did not uncover anything unusual or ominous. As for personnel, they appeared to be predominantly individual replacements with no new NVA or VC units identified. The bottom line of the estimate completed in November 1971 was that 1972 would be "business as usual" without any significant surprises.

By late December, however, new information began to challenge this estimate. The first clues resulted from the analysis of captured documents. A single analyst had detected subtle differences in the tone of South Vietnamese Communist Party documents (which were often filled with hidden messages since the enemy knew some of the documents would be captured) hinting at something big afoot. Shortly thereafter another analyst noted a sudden increase in the flow of personnel into the south. Further projections showed that if this higher level continued through January, it would greatly exceed numbers needed to replace combat losses.

These two indicators cued new collection operations and a renewed analytical effort. Particularly troubling was a photograph of a tank park located in North Vietnam near the South Vietnamese border. Such a concentration of combat vehicles had never been seen that far south. Finally, additional all-source analysis not only verified continued, unprecedented personnel replacements but also uncovered evidence that two new NVA divisions were headed south and would arrive by late February or early March.

The new intelligence caused a complete overhaul of the previous estimate. The new estimate was published in early January 1972. It concluded that the enemy had the capability to initiate a major escalation of the war during the 1972 dry season, beginning any time after the last week in February, using the equivalent of three new divisions and extensive armor forces. Additional hard intelligence indicated that major attacks would occur from the highlands of central Vietnam south to the delta region.

This revised intelligence estimate provided a minimum of 7 weeks' warning of impending enemy actions. This led to another challenge, one routinely faced by the intelligence professional: the more the user is told, the more the user wants to know. The new intelligence requirement was to determine more precisely the date the attack would begin. At the same time, operational commanders used the new intelligence to immediately launch an all-out bombing offensive to impede, weaken, or, if possible, destroy the NVA reinforcements. This had the effect of further complicating the intelligence task, as it was impossible to determine what effect the bombing would have on the enemy's plans. The predicted time for the offensive came and went, and the credibility of the estimate began to be challenged.

The bombing had only delayed the attack. On March 30, the NVA and VC launched an unprecedented offensive. Although not as widespread as the 1968 Tet offensive, it brought the commitment of division-sized regular units accompanied by armor and artillery units, some with weapons that outranged those of the U.S. and South Vietnamese. Even with the advance warning, combat actions were prolonged and intense.

Specific indications of the attack were much clearer for the south and central regions of the country and resulted in more effective defensive operations in those regions. Due primarily to the enemy's ability to better conceal his activities in the border region, intelligence did not adequately detect preparations for offensive operations in the northern region. The estimate did not predict major attacks on the north, and the NVA achieved significant successes in the border provinces.

The accuracy and timeliness of the updated January intelligence estimate was a key factor in ultimately repulsing attacks in the south and central portions of the country. Continuous evaluation of the standing estimate and a willingness on the part of two analysts in particular to challenge conventional wisdom led to this updated assessment. Once initiated, the collection of information from a variety of sources, coupled with in-depth knowledge of the enemy and detailed analysis, provided commanders timely and relevant intelligence they were able to apply to significant advantage.

CONCLUSION

Intelligence strives to build as complete a picture of both the enemy and the area of operations as possible. Such a picture is made up of a variety of factors—the concrete and measurable, the intangible and subjective, the environmental and cultural, the military and political—all of which must be assessed in order to develop the knowledge needed to support the commander's decisionmaking. Building this complete picture requires that we understand and apply the characteristics of good intelligence. Our intelligence picture must be comprehensive; it should combine relevant basic, current, and estimative intelligence from all levels of intelligence. It must include estimates of both capabilities and intentions. Finally, developing an understanding of the situation requires that we be able to distinguish between signals and noise—that we avoid the pitfalls of bias and preconception while interpreting collected information as objectively as possible.

We employ a variety of conceptual tools to help us in achieving our intelligence objectives. Properly defined intelligence requirements are crucial to providing focus to the intelligence effort. The six intelligence functions outline related tasks which, when accomplished, ensure comprehensive intel- ligence support to all phases of operational planning and execution. The intelligence cycle provides a process for the development of intelligence. The six steps in the cycle describe a coordinated sequence of activities which results in the production, dissemination, and utilization of accurate, timely, and relevant intelligence.

The object of the intelligence cycle is not to prescribe a procedure which, if successfully applied, will ensure the quality of the intelligence product. The criterion for good intelligence is not whether the different phases have been properly adhered to and whether an accurate, complete, and polished intelligence product has emerged. Likewise, the discussion of the characteristics of good intelligence is not meant as a checklist. These discussions are meant to emphasize that *the sole criterion for good intelligence is whether it provides sufficient knowledge regarding the environment and an understanding of the enemy's capabilities, limitations, and intentions to effectively support the commander's planning and decisionmaking.*

Chapter 3

Creating Effective Intelligence

"To lack intelligence is to be in the ring blindfolded."[1]

—David M. Shoup

"It is refreshing to see things in their proper order—intelligence driving operations, instead of operations driving intelligence . . . As a consequence, we have been able to maintain a constantly high tempo of productive operations."[2]

—Charles E. Wilhelm

H aving reached a common understanding of the nature of intelligence and having laid out the main elements of intelligence theory, we can describe the characteristics of effective intelligence. How do we create it within in the Marine Corps?

THE CHALLENGE TO INTELLIGENCE

Before discussing our approach to intelligence, it might be helpful to review the challenges that intelligence faces. What obstacles must intelligence overcome, and what must it accomplish?

Our fundamental premise is that intelligence is not knowledge for its own sake, but instead knowledge for the sole purpose of supporting the commander's decisionmaking needs. Knowledge that cannot be acted upon or that commanders choose to ignore is of little value. Consequently, the Marine Corps recognizes that because intelligence is directly connected to action, it is therefore inseparable from command and operations.

Intelligence attempts to reduce uncertainty about a particular hostile situation. Intelligence is fundamentally an imprecise activity, dealing in estimates and probabilities rather than certainties. Intelligence must extract meaning from information that is ambiguous, unclear, and sometimes of unknown reliability. It must synthesize disparate information, attempting to create a coherent picture of the enemy and the area of operations. Intelligence should strive to identify enemy centers of gravity and critical vulnerabilities that commanders can exploit. At the same time, it should provide warning of threats to friendly forces.

Intelligence not only provides knowledge of quantitative factors but also, more importantly, affords insight into intangible aspects of the enemy situation such as his goals and motivations. It should not only describe existing conditions and identify enemy capabilities but should also attempt to estimate likely future conditions and enemy intentions. In addition, it should present that knowledge in the form of coherent, meaningful images that are easily assimilated rather than in the form of accumulated lists or texts.

Intelligence strives to answer three basic sets of questions. The first relates to current capabilities and conditions: "What *can* the enemy do? What conditions currently exist?" The second relates to intentions or future conditions: "What *might* the enemy do? What is the enemy *likely* to do? What is the most dangerous thing he may do? What conditions might or are likely to exist in the future?" And the third—and most important—relates to any implications: "What effect might all of this have on our ability to accomplish the mission?"

In short, intelligence must provide the commander with the *practical knowledge that offers exploitable advantages* over the opposition.

INTELLIGENCE IS A COMMAND RESPONSIBILITY

Creating effective intelligence is an inherent and essential responsibility of command. Intelligence failures are failures of command—just as operations failures are command failures.

The Marine Corps' approach to intelligence demands that commanders be personally involved in the conduct of intelligence activities. The commander must specify requirements and provide guidance to ensure a timely and useful product. Commanders must develop an appreciation for the capabilities and limitations of intelligence. This awareness does not mean just an understanding of concepts and theory, but an understanding of the *practical* capabilities and limitations of intelligence personnel, systems, procedures, and products.

The commander begins the process by providing the guidance and direction necessary for the effective conduct of intelligence operations. The commander establishes the priority intelligence requirements that drive collection, production, and dissemination operations. If a commander does not effectively define and prioritize intelligence requirements, the entire effort may falter. The commander is also required to make the final synthesis of intelligence, arriving at the estimate of the situation which, in turn, serves as the basis for the decision. This is the responsibility of the commander and no one else; while the intelligence officer will provide a recommendation, it is the commander who ultimately determines the meaning of the intelligence provided and how to use it. Additionally, the commander supervises the overall intelligence effort to ensure that the product is timely, relevant, and useful.

Importantly, the commander ensures that intelligence activities support not just the intelligence requirements of the parent unit but the requirements of subordinate commanders as well. The commander should intervene personally when the unit's collection requests or other intelligence support requirements go unsatisfied. Finally, the commander must view the intelligence training of all personnel as a personal command responsibility. This training includes the intelligence awareness of all members of the command as well as the professional development and training of intelligence personnel.

THE COMMAND-INTELLIGENCE CONNECTION

The relationship between the commander and the intelligence officer should be as close as that between the commander and operations officer. Personal involvement in intelligence does not imply that the commander micromanages the intelligence section or assumes the job of the intelligence officer any more than involvement in operations means that the commander takes over as operations officer. Instead, commanders must provide the guidance and supervision necessary for the intelligence officer to support them while at the same time allowing the intelligence officer sufficient latitude for initiative.

In reality, however, the relationship between a commander and intelligence officer is often more difficult to establish and maintain. One reason is that the commander and operations officer usually have more in common in terms of grade, military occupational specialty, age, and experience. In the perspective of some officers, an operations billet is a prelude to command, and many commanders have previously served tours of duty as operations officers in the very same type of unit they now command. Commanders rarely have had the same sort of practical experience in intelligence billets. Consequently, commanders must promote an environment of cooperation, professional support, and mutual respect between themselves and their intelligence officers in which operations and intelligence officers can work together to execute their commanders' intent.

Intelligence requirements are the commander's requirements and not those of the intelligence officer. The commanding officer must provide early and adequate guidance and revise it when necessary. The commander identifies what intelligence is needed while the intelligence officer helps in stating the priority intelligence requirements to meet those needs.

The intelligence officer is not simply a researcher waiting for a task from the commanding officer. *An intelligence officer is an operator who understands the intelligence needs of the unit.* The intelligence officer is knowledgeable of the tactical situation and can anticipate the commander's intelligence requirements based on an understanding of the com- mander's intent and the commander's thought processes. The intelligence officer actively advises the commander on just what intelligence may contribute to success and aggressively carries out intelligence operations to fulfill the intelligence needs of the command.

While the relationship between commander and intelligence officer should be close, they must be careful not to lose their objectivity. The commander and intelligence officer may not always agree on their respective estimates of the hostile situation—this is natural and to be expected. Once the intelligence officer has provided a candid, objective estimate, the commander will assess it and make an independent judgment. Once the commander has made a decision, the intelligence officer must support it fully—while maintaining the detachment necessary to advise the commander if the situation changes or if new evidence indicates that the commander's estimate appears wrong.

During planning and wargaming, the commander will often instruct the intelligence officer to assume the role of an adversary—to attempt to think like the enemy commander—as a means of gaining insights into possible enemy intentions, actions, and reactions. Thus, the intelligence officer often plays the role of devil's advocate, identifying possible ways that the enemy or the environment may interfere with or even defeat friendly plans. In this manner, the intelligence officer helps the commander analyze possibilities and prepare responses to possible developments.

Commanders must exercise caution so as not to judge the effectiveness of intelligence by how accurately it has predicted reality. Commanders must realize that intelligence is the business of estimates, not certainties. A commander harboring unrealistic expectations may discover that the intelligence officer is reluctant to risk any predictions for fear of being wrong. The commander must encourage the intelligence officer to estimate enemy possibilities frankly and not merely provide "safe" facts and figures. Far from being merely a provider of facts and figures—or even a provider of estimates on enemy courses of action—*the intelligence officer should offer trusted advice on friendly courses of action based on knowledge of the hostile situation.*

THE INTELLIGENCE-OPERATIONS CONNECTION

The relationship between intelligence and operations should be as close and direct as that between intelligence and command. In addition to intelligence's influence on the conduct of operations by identifying enemy capabilities and estimat- ing enemy courses of action and possible reactions to friendly courses of action, intelligence also provides important support to operations by helping to identify *friendly critical vulnerabilities* that the enemy may exploit. Thus, the intelligence and operations sections must function in close cooperation throughout the planning and execution of an operation. Neither section can perform effectively without the continuous cooperation of the other.

As in the relationship with the commander, the intelligence officer should cooperate fully with the operations officer but should not develop a personal stake in a particular course of action. Based on knowledge of the hostile situation, the intelligence officer must maintain the freedom to offer advice which disagrees with the advice of the operations staff.[3]

Intelligence officers are themselves operators. The intelligence officer does everything the operations officer does, only in red ink—meaning from the enemy, rather than friendly perspective. The intelligence officer must possess an intimate knowledge of the enemy's methods, capabilities, organizations, and tendencies. At the same time, in order to effectively plan, coordinate, and execute intelligence opera- tions, the intelligence officer requires an in-depth understanding of friendly tactics, capabilities, and intentions.

The relationship between operations and intelligence necessitates mutual support. Just as intelligence identifies opportunities for exploitation through operations, so can operations provide the stimulus for intelligence. Regardless of the primary mission, all operations have an additional object of gaining information about the enemy and the environment. Some operations possess this goal as the primary mission. For example, the objective of a tactical maneuver such as a reconnaissance in force may be to learn more about enemy capabilities and disposition or to solicit the enemy's reaction to a specific situation.

The importance of the intelligence-operations connection is seen in the contrasting approaches to intelligence used by the Luftwaffe and the Royal Air Force (RAF) during the Battle of Britain. The RAF placed intelligence officers throughout the organization down to the squadron level. Thus aircrews received the latest intelligence during tailored pre- mission briefings, and information collected during combat was immediately available for analysis, dissemination, and utilization. In contrast, the Luftwaffe placed intelligence officers at the wing level only. Intelligence support to flying groups and squadrons was marginal throughout the battle, and its lack contributed to the German defeat.[4]

The direct connection between intelligence and operations results in intelligence shaping or even driving the course of operations. Intelligence operations seek to uncover enemy vulnerabilities we can exploit. Opportunities identified by the intelligence effort are used to develop the concept of operations during planning and to initiate specific tactical actions during execution. Effective intelligence guides us towards enemy weaknesses rather than forcing us to operate against an enemy strength.

The invasion of Tinian during World War II provides an illustration of how intelligence shapes operations. Initial intelligence studies of Tinian identified only one suitable landing area for the amphibious assault. This area was located immediately in front of the island's major settlement, Tinian Town, and was heavily defended by the Japanese. The studies noted the existence of two small inlets on the northern tip of the island but discounted their suitability for a major landing (see figure 5). As planning progressed, new intelligence identified major disadvantages in attacking across the Tinian Town beaches. At the same time, additional studies indicated that a landing on the undefended northern beaches was a viable option. Preassault reconnaissance confirmed the suitability of these beaches. The concept of operations called for regimental-sized landings to be conducted on two small northern beaches (White 1 and White 2) that totaled in width only about 220 yards. The main landing would be supported by an amphibious demonstration conducted near Tinian Town.

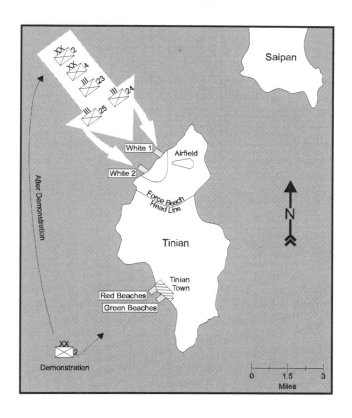

Figure 5.

The amphibious assault achieved complete tactical surprise. Landing against minimal opposition on the northern beaches, the entire 4th Marine Division was established ashore by the end of the first day. The island was secured in 9 days with minimal casualties suffered by the landing force. The Tinian operation was described by Admiral Raymond Spruance as, "probably the most brilliantly conceived and executed amphibious operation in World War II." Intelligence contributed significantly to the success of this operation, providing commanders with knowledge of a critical vulnerability—the undefended northern beaches—which they exploited to achieve success.[5]

INTELLIGENCE AS A TEAM EFFORT

Intelligence is the commander's responsibility and the intelligence officer's primary duty, but it is also definitely the concern of every Marine. All Marines in the command contribute in one way or another to the intelligence effort. Nearly every Marine, regardless of occupational specialty, has occasion to observe significant facts about the enemy or the environment. Units in contact with the enemy are a particularly valuable source of information. All Marines should consider themselves as potential intelligence sources and, equally important, as counterintelligence assets. Everyone on the battlefield should be alert for important information and bring that information to the attention of the person who needs it by the most direct and expeditious means available.

INTELLIGENCE IS A PRODUCT, NOT A PROVISION

Intelligence is something generated through our own efforts, rather than something provided by some outside source. While we may say that in principle we should have ready access to external sources like satellite imagery, basing our actions on the timely availability of such information is dangerous. Commanders should aim, to the greatest extent possible, to become self-sufficient in satisfying their own intelligence requirements. This approach is particularly important once an operation has commenced. Before operations begin, intelligence from higher echelons may appear to be available in unlimited quantities. However, once execution starts, our organic intelligence and reconnaissance assets generally provide the most reliable and responsive support to Marine units. Marines cannot forget that intelligence is the result of solid headwork and legwork, and it is not provided from some omniscient source of knowledge. Requirements for critical intelligence should be satisfied through organic means whenever possible.

A BALANCED APPROACH

The approach of the Marine Corps to intelligence calls for balance in a number of areas. First is the capability to gather information from a variety of sources. Each source provides a different type of information. These different sources can compensate for, complement, and confirm one another. Depending on the situation, certain sources will be more valuable than others. Which source we most depend upon in a particular situation is less a matter of our own preference than a matter of the nature and sophistication of the enemy.

Next, balance means that commanders emphasize equally all phases of the intelligence cycle. For example, an overemphasis on collection may result in an overload of information that overwhelms processing and production capacity, thus preventing rapid dissemination. Balance also means that commanders emphasize the development of both classes of in- telligence—descriptive and estimative. Balance requires that intelligence personnel work at uncovering both the enemy's capabilities and the enemy's intentions. Balance means that we take into account both quantitative factors—such as numbers, locations, equipment specifications—and qualitative factors—morale, motives, leadership, and cultural values.

Finally, our approach to intelligence should achieve balance in its support to commanders at all levels. At any particular level, the intelligence officer's first duty is to serve the commander's intelligence requirements. However, since questions about the enemy situation and area of operations are practically limitless, an intelligence section can easily spend all its time satisfying intelligence requirements of its own staff or higher headquarters—to the neglect and detriment of subordinate commanders' intelligence needs. Commanders must provide the necessary guidance to ensure that balance is achieved.

Intelligence personnel must remain conscious of the intelligence requirements of all elements of the force with the objective of creating satisfactory intelligence for all supported commanders.

FOCUSING THE INTELLIGENCE EFFORT

Focus, as embodied in the concept of main effort, is central to maneuver warfare. It is particularly critical for intelligence since possible questions about the enemy situation are nearly infinite, while intelligence assets are limited. Commanders must concentrate intelligence operations on those critical requirements upon which mission success depends and prioritize accordingly. *The intelligence effort must support the main effort.* In fact, intelligence is responsible for identifying the enemy's centers of gravity and critical vulnerabilities that are used to determine the main effort.

Focus is a product of direction, which means it is a function of command. The commander provides focus to the intelligence effort by prioritizing intelligence requirements. These requirements establish priorities for all intelligence activities—not only for collection but also for processing, production, and dissemination. The earlier the commander establishes this guidance, the more focused, timely, and complete the final intelligence product will be.

GENERATING TEMPO THROUGH INTELLIGENCE

Tempo is central to maneuver warfare and our command and control doctrine. We seek to generate a higher operating tempo than that of the enemy. Effective intelligence supports this accelerated tempo. We help create this rapid tempo through a variety of techniques.

First, we generate tempo through *prioritization*. We establish a limited number of priority intelligence requirements that are understood clearly throughout the force. Collection, processing, production, and dissemination operations are conducted in accordance with these priorities. By concentrating on the truly essential requirements, we avoid diluting intelligence operations and clogging dissemination channels with nonessential intelligence.

Another way we use intelligence to accelerate tempo is through *decentralization*—establishing command relationships or task-organizing intelligence assets to directly support subordinate commanders. Decentralizing intelligence re- sources applies not only to collection assets but to production and dissemination assets as well. Decentralization provides subordinate elements with the intelligence resources needed to recognize and exploit opportunities as they arise in the battlespace. It also helps ensure intelligence products are tailored to the requirements of commanders at lower echelons. However, decentralization does not mean that intelligence assets will be fully dispersed throughout the force or that each unit will have an equal share of the available intelligence units, systems, or personnel. Since intelligence assets are limited, it is virtually impossible to provide each unit with all the intelligence capabilities its commander may desire. Assets will be allocated based on the commander's intent, the designation of the main effort, and the priority intelligence requirements.

The third technique by which we generate intelligence tempo entails a conscious command decision *to disseminate certain information before it has been fully integrated, analyzed, evaluated, and interpreted*—in other words, before it becomes a comprehensive intelligence product. This approach recognizes that at times a piece of information may be so critical and time-sensitive that it should be disseminated immediately with minimal evaluation and analysis. In a sense, this amounts to decentralizing intelligence production by requiring subordinate units to perform immediate intelligence production. Immediate production rapidly identifies, evaluates, and disseminates intelligence that may have an impact on ongoing operations in order to exploit opportunities and generate tempo. For example, the commander may establish criteria that require the immediate dissemination of any reporting on certain critical enemy targets. The dilemma, which we must resolve on a case-by-case basis, is between the desire to provide as complete and accurate an intelligence product as possible and the requirement to support the urgency of tactical decisionmaking.

Accessibility is a fourth mechanism by which we may accelerate tempo. Accessibility increases tempo by making intelligence available to commanders for use in decision- making. Here we make another conscious command decision—in this case to make intelligence more accessible to users by *minimizing security restrictions or by relying more on open sources.* We can do this by "sanitizing" classified intelligence to protect sources without materially decreasing the value of the intelligence. More important, we should make a conscious effort to ensure that intelligence is classified only to the minimum degree essential to the interests of security.

A fifth way of generating tempo is by ensuring that intelligence products take the *form* most readily understandable by users. This generally means that *intelligence should be presented as meaningful images, rather than reports or lists which require more time to assimilate.* For example, displaying a possible enemy course of action in a graphic with supporting text annotated on the graphic is generally more useful than providing only a text report.

Finally, we can enhance tempo through effective *information management*—taking advantage of all available communication channels and means for disseminating intelligence. Intelligence, like any other information product, flows not only through established hierarchical channels but also by alarm channels, flowing laterally and diagonally as well as vertically (see figure 6). In other words, rather than simply forwarding information or intelligence via standard channels, we must ask ourselves, "Who really needs this information most?" and transmit that information by the most direct and readily accessible means.

The ability to generate tempo through intelligence was vividly demonstrated in a series of combat actions during the early years of the Vietnam war. A small number of documents recovered from a Viet Cong commander killed by a Marine am-

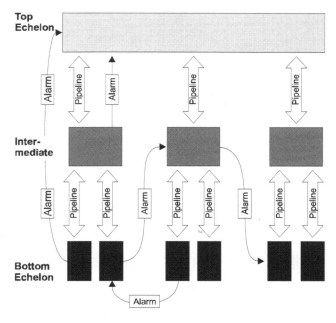

Figure 6.

bush patrol identified a likely enemy training base. Using this intelligence, the next day a Marine combined arms assault surprised and effectively destroyed five enemy companies. Immediate searches of the area led to additional intelligence locating another enemy battalion, which was also quickly attacked, causing heavy personnel and material losses. Follow-on all-source intelligence analysis of both engagements swiftly identified the most likely infiltration routes used by these enemy units, allowing a Marine infantry company a few nights later to successfully ambush a reinforced enemy battalion. A few captured documents combined with rapid dissemination and utilization of the resulting intelligence led directly to a series of successful tactical actions.[6]

INTELLIGENCE EDUCATION AND TRAINING

Intelligence education and training are a command responsibility. Professional development programs must give all Marines an understanding of the capabilities and limitations of intelligence as well as the employment of intelligence assets. Education and training should likewise provide intelligence personnel with an in-depth understanding of operations so that they may better support operations with intelligence. Moreover, education and training programs should seek to strengthen the relationship between intelligence officers and commanders by increasing their mutual understanding.

Commanders must demonstrate personal involvement in intelligence training. They must dedicate adequate training time to it and ensure that intelligence is realistically integrated and balanced with other warfighting activities. The commander is responsible for ensuring that all the unit's Marines have a basic understanding of the threat and the environment in likely areas of deployment. Classroom instruction, professional reading, discussion groups, and use of wargames with realistic scenarios and threat forces are ways to build such knowledge.

Exercises must be used to reinforce and increase the intelligence awareness of the unit. Exercises should incorporate realistic intelligence to the maximum extent possible. This provides participants with the opportunity to identify their intelligence requirements, allows them to see how intel- ligence is collected, produced, and disseminated, and exposes them to the type and quantity of intelligence support they can expect to receive during actual operations.

The value of incorporating realistic intelligence into exercises was demonstrated during Operation Praying Mantis in 1988. The commander of a Special Purpose Marine Air-Ground Task Force (MAGTF) based his training scenarios on actual intelligence studies of potential raid sites in the Persian Gulf. For the execution of the operation, the MAGTF was directed to attack the "Sassan" gas-oil platform, a target the

MAGTF had used in its training exercise the week before. The use of realistic intelligence during training gave the MAGTF commander the necessary background knowledge and situational awareness to rapidly complete the plan.[7]

Opposed, free-play exercises are especially valuable, providing the opportunity to conduct intelligence operations in realistic conditions. To use an intelligence staff to create an exercise scenario, with all the pertinent intelligence already generated in advance, is a misuse of assets. Such a scheme robs an exercise of all-important realism in the development and use of intelligence to support decisionmaking. Within the practical limitations of available resources, "scripted" exercise intelligence should be minimized in favor of intelligence generated during the actual exercise.

In the training and education of intelligence personnel, we seek to achieve a balance between specialization and generalization. *Intelligence officers must possess a broad operational orientation*—an understanding of just how intelligence supports operations in general terms—while also developing the specialized skills required by many intelligence disciplines. We should nurture intelligence officers who can synthesize as well as analyze—who can answer the "So what?" question. Finally, we should stress the importance of foreign area and foreign language training in order to build our understanding of potential enemies and operating environments.

A CASE STUDY: SOMALIA 1992–1993

In December 1992, lead elements of the 15th Marine Expeditionary Unit landed in Mogadishu, Somalia, initiating Operation Restore Hope, a multinational humanitarian assistance operation. Remaining elements of Marine Forces (MARFOR) Somalia followed shortly thereafter. MARFOR intelligence operations illustrate the importance of a commander's involvement in the intelligence effort and of close coordination between intelligence and operations.[8]

The intelligence situation at the start of Operation Restore Hope was typical of what can be anticipated for most military operations other than war, particularly from the tactical perspective: outdated basic intelligence, sketchy current intelligence regarding the order of battle, capabilities, intentions, and vulnerabilities of potential threat forces, and limited understanding of possible reactions from either the civilian populace or the many nongovernmental organizations long operating in the country.

From the beginning, the commander ensured that MARFOR intelligence and operations elements worked as a team. The commander set the direction for MARFOR intelligence operations by focusing the collection and production efforts. During the initial stabilization phase of operations, intelligence requirements were focused on the organization and leadership of the

Somalian clans, boundaries between the clans, and the locations of meeting places, weapons caches, and arms markets. Both intelligence and operations personnel worked to acquire information and develop understanding of the nongovernmental organizations, the status of the local infrastructure, and the cultural aspects of the local population. During the subsequent normalization phase of the operation, intelligence priorities shifted to requirements in support of the civil affairs effort: preserving freedom of movement and commerce throughout the country, determining the effectiveness of civilian authorities, and estimating the attitudes of the clans and the average Somali to U.S. and U.N. efforts.

Collection operations reflected the unique challenges of the humanitarian assistance mission. A considerable amount of information was acquired from foot, motorized, and mechanized patrols. Helicopter visual reconnaissance missions and postmisson debriefs typically provided timely confirmation of information acquired during patrols. The MARFOR's principal human intelligence resources, its counterintelligence and interrogator-translator teams, were exceptionally effective in this environment.

Counterintelligence specialists and interrogator-translators were routinely attached to or placed in direct support of units down to battalion and regimental level. Their immediate availability and integration into unit intelligence collection and other operational activities allowed intelligence officers to rapidly develop pertinent tactical intelligence. In most instances, intelligence was immediately provided to and acted upon by MARFOR operational elements. When more complex targets were identified, intelligence was used to plan and execute sophisticated direct action missions. The effective development and use of intelligence led to the capture of hundreds of weapons and tons of ammunition and supplies. Intelligence contributed directly to the establishment of a secure environment for the conduct of relief activities.

The MARFOR commander characterized human intelligence operations as "providing in-by-nine, out-by-five service on priority intelligence requirements. As a consequence, we have been able to maintain a constantly high tempo of productive operations. The key word here is productive. Patrols, checkpoints, and direct action missions have, for the most part, been directed against clearly defined targets—there have been remarkably few dry holes."[9]

CONCLUSION

The Marine Corps' philosophy of intelligence recognizes that intelligence is an inherent responsibility of command. The commander's direct involvement is required to provide appropriate guidance to the intelligence effort and ensure the full utilization of the intelligence product. Our philosophy also acknowledges that intelligence is inseparable from operations and that effective intelligence shapes or even drives operations. Without close and continuous cooperation, neither intelligence nor operations can function effectively.

Our intelligence philosophy relies on a variety of sources, does not emphasize one phase of intelligence activity at the expense of another, and provides support to all levels of the force. This approach recognizes the importance of qualitative as well as quantitative information requirements. It focuses on priority intelligence requirements, seeking to avoid diffusion of effort. The Marine Corps' intelligence philosophy acknowledges the importance of tempo and uses effective intelligence operations to develop and maintain tempo. Finally, this approach to intelligence recognizes that the obtaining of useful information about an adversary is a team effort and requires that all Marines see themselves as intelligence and counterintelligence resources contributing actively to the intelligence effort.

The Nature of Intelligence

1.   Sun Tzu, *The Art of War*, trans. Samuel B. Griffith (New York: Oxford University Press, 1963) p. 129.

2.   Stedman Chandler and Robert W. Robb, *Front-Line Intelligence* (Washington, D.C.: Infantry Journal Press, 1946) p. 7. Available to Marine units as FMFRP 12-16, *Front-Line Intelligence*.

3.   Erwin Rommel, *The Rommel Papers,* ed. B. H. Liddell Hart, trans. Paul Findlay (New York: Harcourt, Brace, 1953) p. 122.

4.   The surrounding environment includes weather, terrain, transportation network, local population, and any other factors needed to describe a potential area of employment.

5.   Sherman Kent, *Strategic Intelligence for American World Policy* (Princeton, NJ: Princeton University Press, 1966) p. 3.

6.   For a detailed explanation of the distinction between data, information, knowledge, and understanding, see MCDP 6, *Command and Control* (October 1996) pp. 66–71.

7.   Wayne M. Hall, "Intelligence Analysis in the 21st Century," *Military Intelligence* (January-March 1992) p. 9 and Richard K. Betts, "Analysis, War and Decision: Why Intelligence Failures Are Inevitable," *Power, Strategy, and Security*, ed. Klaus Knorr (Princeton, NJ: Princeton University Press, 1983) p. 211.

8.   Kent, *Strategic Intelligence for American World Policy*,
p. 180.

9.   Nancy Gibbs, "A Show of a Strength," *Time* (October 1994) pp. 34–38.

10.   Sherman Kent, "Estimates and Influence," *Foreign Service Journal* (April 1969) p. 17.

11.   Principal sources used in this case study: *Conduct of the Persian Gulf War: Final Report to Congress* (Washington, D.C.: Department of Defense, April 1992); Col Charles J. Quilter, II, *U.S. Marines in the Persian Gulf, 1990–1991: With the I Marine Expeditionary Force in Desert Shield and Desert Storm* (Washington, D.C.: Headquarters, U.S. Marine Corps, History and Museums Division, 1993); LtCol Charles H. Cureton, *U.S. Marines in the Persian Gulf, 1990–1991: With the 1st Marine Division in Desert Shield and Desert Storm* (Washington, D.C.: Headquarters, U.S. Marine Corps, History and Museums Division, 1993) pp. 46; Dennis P. Mroczkowski, *U.S. Marines in the Persian Gulf, 1990–1991: With the 2d Marine Division in Desert Shield and Desert Storm* (Washington, D.C.: Headquarters, U.S. Marine Corps, History and Museums Division, 1993); Interview with Maj-Gen J. M. Myatt, USMC, "The 1st Marine Division in the Attack," *Proceedings* (November 1991) pp. 71–76; Interview 4 Nov 96 with LtCol David Hurley and Mr. Michael H. Decker conducted by Maj Emile Sander (during Operation Desert Shield/Desert Storm, LtCol Hurley served as the officer in charge of the I MEF MAGTF all-source fusion center, while then Capt Decker served as the MAGTF all-source fusion center, Chief Intelligence Analyst).

Intelligence Theory

1.   Carl von Clausewitz, *On War*, trans. and ed. Michael Howard and Peter Paret (Princeton, NJ: Princeton University Press, 1984) p. 117.

2.   Gen Carl E. Mundy, Jr., "Reflections on the Corps: Some Thoughts on Expeditionary Warfare," *Marine Corps Gazette* (March 1995) p. 27.

3.   Maj David L. Shelton, "Intelligence Lessons Known and Revealed During Operation Restore Hope Somalia," *Marine Corps Gazette* (February 1995) pp. 37–40; Capt David Rababy, "Intelligence Support During a Humanitarian Mission," *Marine Corps Gazette* (February 1995) pp. 40–42; LtCol David Shelton, "Intelligence Lessons Known and Revealed During Operation United Shield in Somalia," unpublished manuscript.

4.   Richard K. Betts, *Surprise Attack: Lessons for Defense Planning* (Washington, D.C.: Brookings Institution, 1982) pp. 56–62. See also Edwin P. Hoyt, *The Bloody Road to Panmunjorn* (New York: Military Heritage Press, 1985) pp. 17, 31, 45–47, 54–59.

5.   U.S. Department of Defense, *Report of the DOD Commission on Beirut International Airport Terrorist Act, October 23, 1983* (Washington, D.C.: December 20, 1983) pp. 64–66.

6.   The senior military intelligence officer in Vietnam at the time of the Tet offensive later declared, "Even had I known exactly what was to take place, it was so preposterous that I probably would have been unable to sell it."   Jack Shulimson, *TET–1968* (New York: Bantam Books) p. 45. See also Don Oberdorfer, *Tet!* (New York: Da Capo Press, 1984) pp. 117–121.

7.   For a full discussion of signals and noise, see Roberta Wohlstetter, "Cuba and Pearl Harbor: Hindsight and Foresight," *Foreign Affairs* (July 1965) pp. 691–707.

8.   Clausewitz, p. 101.   •

9.   Betts, *Surprise Attack: Lessons for Defense Planning*, pp. 68–80.

10.   Ephraim Kam, *Surprise Attack: The Victim's Perspective* (Cambridge, MA: Harvard University Press, 1988).

11.   The terms *priority intelligence requirement* (PIR) and *intelligence requirement* (IR) replace the terms *essential elements of information* (EEI) and *other intelligence requirements* (OIR) to correspond to joint doctrinal terminology. Priority intelligence requirements are a subset of *commander's critical information requirements* (CCIRs), a term used to describe the information and intelligence requirements a commander deems critical to mission success. CCIRs are discussed in FMFM 6-1, *The Marine Division* (to be redesignated as MCWP 3-11.1).

12. Interview with LtCol Dwight Trafton, Intelligence Officer, 24th Marine Expeditionary Unit, December 30, 1996, conducted by LtCol J. D. Williams.

13. Specific security guidelines—particularly restrictions and information classification guidance—are often dictated by higher echelon commanders in pertinent directives, operation orders, or classification guides.

14. For information on the use of signals intelligence during World War II, see David Kahn, *The Codebreakers: The Story of Secret Writing* (New York: New American Library, 1973); F. W. Winterbotham, *The Ultra Secret* (New York, Harper & Row, 1974); Ronald Lewin, *Ultra Goes to War: The First Account of World War II's Greatest Secret Based on Official Documents* (New York: McGraw-Hill, 1978). For a discussion of the operation against Admiral Yamamoto, see Kahn, pp. 332–338.

15. The principal source used in this case study is Col Peter F. C. Armstrong, USMC (Ret.), "Capabilities and Intentions," *Marine Corps Gazette* (September 1986) pp. 38–47. Additional information was taken from Stanley Karnow, *Vietnam: A History* (New York: Penguin Books, 1984) pp. 639–641.

16. The southwest monsoon season ran from mid-March to mid-September, flooding much of the South Vietnamese delta while turning the Laos and Vietnamese highlands into a quagmire nearly impassable to heavy vehicles. As a result, major military operations could not begin until the ground had dried, normally in December.

Creating Effective Intelligence

1.   Gen David M. Shoup, USMC, in Robert Debs Heinl, Jr., *Dictionary of Military and Naval Quotations* (Annapolis, MD: United States Naval Institute, 1966) p. 161. Gen Shoup was the 22nd Commandant of the Marine Corps.

2.   MajGen Charles E. Wilhelm, Commander, Marine Forces, Somalia, Situation Report for January 24, 1993.

3.   Kent, *Strategic Intelligence for American World Policy*,
p. 198.

4.   LtCol David Ingram, " 'Fighting Smart' with ACE Intelligence," *Marine Corps Gazette* (May 1989) pp. 38–40. See also Len Deighton, *Fighter: The True Story of the Battle of Britain* (New York: Harper Paperbacks, 1994).

5.   Maj Carl Hoffman, *The Seizure of Tinian* (Washington, D.C.: Headquarters, U.S. Marine Corps, Historical Division, 1951).

6.   FMFRP 12-40, *Professional Knowledge Gained from Operational Experience in Vietnam, 1965–1966* (September 1991) pp. 139-141.

7.   Col William M. Rakow, "Marines in the Gulf–1988," *Marine Corps Gazette* (December 1988) pp. 62–68.